ONE LIFE, ONE PLANET

ONE KOALA IN FIFTY THOUSAND

SARAH RIDLEY AND VIVIAN MINEKER

WAYLAND

Natural History Museum

First published in Great Britain in 2024
by Wayland
Copyright © Hodder and Stoughton, 2024

All rights reserved

Editor: Sarah Peutrill
Designer: Peter Scoulding

ISBN (HB): 978 1 5263 2294 4
ISBN (PB): 978 1 5263 2293 7

Printed and bound in China
Wayland, an imprint of
Hachette Children's Group
Part of Hodder and Stoughton
Carmelite House
50 Victoria Embankment
London EC4Y 0DZ
An Hachette UK Company

www.hachette.co.uk
www.hachettechildrens.co.uk

Additional picture credits:
Page 29 (globe) Spreadthesign/Shutterstock
Page 30 (top): rickyd/Shutterstock
Page 30 (bottom): Frank Martins/Shutterstock

Every attempt has been made to clear copyright. Should there be any inadvertent omission please apply to the publisher for rectification.

ONE LIFE, ONE PLANET

ONE KOALA IN FIFTY THOUSAND

SARAH RIDLEY AND VIVIAN MINEKER

ONE hungry koala clings to the trunk of a eucalyptus tree. She uses her hands to reach some tasty leaves. Is this one good to eat? She sniffs it carefully and pops it into her mouth.

The koala starts to chew – and chew – **AND CHEW**. The leaves are very hard. She chews more than 16,000 times a day to break them down!

The koala lives in a eucalyptus forest on the east side of Australia. Wild koalas live nowhere else on Earth.

The koala sniffs some leaves before popping them into her mouth. She is fussy about which leaves she eats. She hardly ever needs a drink as her body gets most of the water it needs from the leaves.

The leaves are all she eats but they contain poison. When she sniffs them with her big nose, she can tell which ones contain less poison – and she chooses to eat those. Inside her body, tiny living things called bacteria help to break down the poison and the leaves. It takes a long time.

The koala spots some tasty leaves on another tree. Down she climbs. Sharp claws on her hands and feet, as well as gripping fingers and toes, make this easy.

Now she's on the ground, she watches out for danger. There is a dingo over there and a goanna halfway up another tree. They will eat her up if they get a chance. Soon she will climb into the treetops again where she will be safe from most predators.

That was all very tiring! The koala falls asleep. Thick fur on her bony bottom acts like a cushion, making her comfortable in the fork of a tree.

ZZZZZZZZZZZZ. She sleeps for between **TWENTY** and **TWENTY-TWO** hours a day, which is almost all the time! She needs to sleep so much because the leaves she eats have few nutrients, and it takes a lot of energy for her body to digest them.

The koala is still resting. What's that bird doing on my back, she wonders? It is plucking out some of her fur and flying off with it to line its nest.

The bird is a brown-headed honeyeater. It is building a cup-shaped nest from strips of bark, grasses, cobwebs and fur. The bird likes to use koala fur to line its nest as the fur keeps its eggs and chicks warm in winter and cool in summer.

Along comes another honeyeater to collect fur from the koala.

When she is about three years old, the koala becomes a mum. She gives birth to a tiny baby called a joey. It's only the size of a grape. Pink and blind, with no fur or ears, the joey crawls her way up her mother's body and into her pouch. There she stays, drinking her mother's milk from a teat inside the pouch.

After about seven months, the joey is bigger and looks like her mum. As well as milk, the baby eats a special sort of poo, called pap, made by her mother. This brings useful bacteria from her mother's body into the baby's body so that she can digest eucalyptus leaves too.

SWOOSH. That was close. A wedge-tailed eagle almost grabbed the joey. She would have made a tasty snack for the eagle and her chicks. Today the eagle will have to carry on hunting. She mostly hunts rabbits and hares as well as feeding on dead animals.

The joey rides on her mother's back to travel through the trees. She spends less and less time in the pouch until she stops returning there at all. Slowly the joey learns which leaves to eat and how to stay safe in the forest.

The koala and her joey keep themselves to themselves. The mother is only interested in other koalas when she wants to find a mate. Now her baby is almost one, she will look for a mate soon.

The mother and her joey stick to their own patch of forest, made up of about a hundred trees. Of course, they are not the only koalas in this forest. And there are other eucalyptus forests where koalas live. Koalas are hard to count but people think there are between about **FIFTY THOUSAND AND ONE HUNDRED THOUSAND** wild koalas in Australia.

All those **THOUSANDS** of koalas do what almost all animals do - poo. In fact, koalas poo even when they are asleep!

Each koala produces up to 360 small poos a day. If it rains, some of the poos will break up and wash into the soil or be buried by animals when they are digging.

Other poos provide food and shelter for insects. Dung beetles sniff out the poos when they drop to the ground and take them away to lay eggs in them. That's if a koala scat moth doesn't get there first. It lays its eggs in koala poos. Each caterpillar feeds on the poo until it changes into a moth and pulls itself out of its koala poo home.

All across eucalyptus forests, koala poos break down into tiny bits, adding nutrients to the soil. The roots of eucalyptus trees and seedlings soak up the nutrients and use them to make their own food to grow and stay healthy.

Like all plants, eucalyptus trees help our planet by releasing the oxygen we breathe and capturing carbon dioxide from the air. This is good for planet Earth as it means less carbon dioxide reaches Earth's atmosphere where it warms the planet.

Trees absorb carbon dioxide during the daytime

Trees release oxygen during the daytime

The eucalyptus trees are also home to all sorts of animals, including birds.

Rainbow lorikeets suck up nectar from eucalyptus flowers, moving pollen from flower to flower.

Black-chinned honeyeaters feed on eucalyptus nectar and pollinate the flowers at the same time, so the flowers can make fruit and seeds.

Long-billed corellas nest in hollows in the trunks of older eucalyptus trees.

The roots of grasses, small trees and bushes growing on the forest floor soak up koala poo nutrients too. That includes golden wattle plants. These small trees are some of the first plants to regrow after a bushfire, helping the forest to recover.

When a golden wattle bursts into flower, bees and other insects flock to its flowers to eat or collect pollen.

Butterflies lay their eggs on the wattle leaves so their caterpillars can munch on them.

Small birds visit the plant in search of sweet nectar. They move pollen from flower to flower, so the flowers can make fruit and seeds.

Colourful superb fairywrens perch in the branches. They pluck ants from the wattle and will eat its seeds.

Without **THOUSANDS** of koalas and their poo adding nutrients to the soil, there would be fewer plants in the eucalyptus forests. Fewer plants would mean fewer of the **MILLIONS** of insects, birds and mammals that live in the forests. It is amazing to think how important koalas are when they spend so much of every day asleep!

Wombats make dens underground or around the roots of eucalyptus trees.

Sleepy possums live in eucalyptus forests and mostly eat eucalyptus leaves.

Kangaroos pass through the forest, eating grasses. They make tracks through the grasses which koalas use to move from tree to tree.

So far people have discovered almost **TWO MILLION** different sorts, or species, of living thing but it's possible that there are as many as **EIGHT MILLION** different species on Earth. Each species is linked to others in all kinds of ways. This makes each one important.

Together all the different species create BIODIVERSITY, the great variety of life on Earth. Biodiversity keeps our planet, and all that lives on it, healthy and safe for the future.

The lives of THOUSANDS of koalas are linked to MILLIONS of plants and animals in eucalyptus forests. Everything that lives in the forests is linked to other living things in habitats close by.

HOW TO HELP KOALAS

The Australian government names koalas as endangered animals. That means they are at risk of dying out completely. This is because their forest home is being chopped down to clear the land for roads, buildings or farmland. Also, climate change is causing more and more bushfires which burn large areas of forest to the ground. There are also diseases which spread from koala to koala.

You can help koalas by learning about them and passing on what you know to others, so that many more people learn how important it is to save koalas and their forest homes.

You can also join organisations such as the World Wide Fund for Nature (WWF) and charities which work to protect wildlife in Australia, including the Australian Conservation Foundation, the Australian Koala Foundation, the Koala Clancy Foundation and the International Fund for Animal Welfare (IFAW).

KOALA FACTS

- Koalas have lived on Earth for millions of years.

- Huge ears help a koala hear very well. Its big nose gives it a great sense of smell.

- Koalas are more active at night. They spend most of their lives sleeping or resting.

- People often call koalas 'koala bears' but they are marsupials. Marsupials are mammals which give birth to tiny young, often keeping them safe in pouches for the first months of life.

- Out of over 800 different sorts of eucalyptus tree, koalas only like to eat the leaves of about 50 of them.

- Koalas have waterproof fur. Rainwater just runs off them.

- Male koalas are much bigger than females and they attract a mate by making low noises, a bit like burps.

- A koala's pouch opens towards its back legs rather than towards its head.

- Forests used to cover much more of Australia. They were home to millions of koalas.

- Predators of koalas include foxes, dogs, cats, snakes and goannas. Owls and eagles are strong enough to hunt baby koalas.

- Koalas bring millions of tourists to Australia every year, keen to see these amazing animals. By bringing attention to the importance of their forest homes, koalas help the other plants and animals that live in their forest homes.

GLOSSARY

ATMOSPHERE the gases surrounding Earth
BIODIVERSITY the great variety of life on Earth
BUSHFIRE a fire spreading fast through a wild area
CARBON DIOXIDE a gas plants use to make their food; a gas breathed out by animals
CLIMATE CHANGE changes in the normal weather patterns around the world
DIGEST break down food inside the body
GOANNA a large lizard
MATE the partner of an animal
NECTAR a sweet liquid made inside flowers
NUTRIENT a substance that helps animals and plants to live and grow
OXYGEN a gas that all animals need to breathe to stay alive and which is released by plants
POISON something that causes harm or even death if it gets into the body
POLLEN dust made by the male part of a flower
POLLINATE to move pollen from the male part of a flower to the female part of another flower, so that the flower can make fruit and seeds
PREDATOR an animal that hunts other animals
SEED the small part of a plant from which a new plant can grow
SEEDLING a young tree or plant
SPECIES a kind of living thing, such as a koala

INDEX

biodiversity 28–29
birds 13, 23, 25–26
   honeyeaters 13, 23, 25
   long-billed corellas 23
   rainbow lorikeets 23
   superb fairywrens 25
   wedge-tailed eagles 17
bushfires 24

carbon dioxide 22

dingoes 8

Earth's atmosphere 22
eucalyptus trees 5–6, 8, 10, 15, 17–18, 22–23, 26–28

goannas 8
golden wattles 24–25

insects 21, 24–26
   bees 24
   butterflies 25
   dung beetles 21
   koala scat moths 21

kangaroos 27
koalas 5–6 and *throughout*
   eating 5–10, 15, 17
   fur 10, 13, 15
   joeys 15–18
   poo 15, 21–22, 24, 26
   predators 8, 17
   sleeping 10–13, 21, 26

oxygen 22

possums 27

wombats 26